AF320499

Veronika Lushchik

Simulation modelling in industrial biotechnology

AF320499

Veronika Lushchik

Simulation modelling in industrial biotechnology

ScienciaScripts

Imprint

Any brand names and product names mentioned in this book are subject to trademark, brand or patent protection and are trademarks or registered trademarks of their respective holders. The use of brand names, product names, common names, trade names, product descriptions etc. even without a particular marking in this work is in no way to be construed to mean that such names may be regarded as unrestricted in respect of trademark and brand protection legislation and could thus be used by anyone.

Cover image: www.ingimage.com

This book is a translation from the original published under ISBN 978-3-659-71068-1.

Publisher:
Sciencia Scripts
is a trademark of
Dodo Books Indian Ocean Ltd. and OmniScriptum S.R.L publishing group

120 High Road, East Finchley, London, N2 9ED, United Kingdom
Str. Armeneasca 28/1, office 1, Chisinau MD-2012, Republic of Moldova, Europe
Printed at: see last page
ISBN: 978-620-7-98579-1

Copyright © Veronika Lushchik
Copyright © 2024 Dodo Books Indian Ocean Ltd. and OmniScriptum S.R.L publishing group

TABLE OF CONTENTS:

INTRODUCTION

Such branches of national economy as ecology, biotechnology, medical, food and microbiological industries are very important for the present and future of mankind. In these industries the object investigated in this paper is widely used.

At present, much attention is paid to the study of various methods of controlling the evolutionary processes of various biological communities. In he interest in tasks of this kind is connected with the emergence of new practically important tasks related to problems of ecology, environmental protection, development of new drugs, and also with the development of technologies for artificial biosynthesis of microorganism populations.

There is a need to develop new technologies that allow saving energy and material resources. Along with the development of energy- and resource-saving technologies, various methods of active artificial impact on the environment play an increasingly important role, allowing to effectively influence the conditions of development of individual species (populations) of various biological communities. In this connection, the development of methods for population management is of great importance. Taking into account the general trend of development of economical technologies, the development of methods for optimal management of individual populations and biological communities is of the greatest interest. It should be noted that although recently there have been quite a large number of publications devoted to the problems of control (including optimal control) of population sizes described by deterministic models of evolutionary type, nevertheless, a wide range of practically important problems of analysis and control of biological communities whose behaviour is described by deterministic and nonlinear stochastic differential equations remains poorly studied.

The problems of population size management have become particularly important as a result of the development of methods of artificial industrial biosynthesis based on the use of microbial communities.

The cultivation of useful biomass of microorganisms (bacteria) is carried out in special devices called chemostats. The purpose of chemostat control is usually to ensure the operating mode corresponding to maximum productivity and maximum yield of useful microbial biomass. Equally important are also the problems of analysing the population dynamics in a chemostat under different methods of controlling the input flow of nutrient substrate. Thus, the solution of specific problems of population control is of great importance in the development of resource-saving technologies in industrial microbiology, as well as in the development of optimal strategies for the utilisation of natural resources for economic purposes.

All populations are characterised by such processes as birth and death of individuals, their dispersal, predation, competition, parasitism, spread of diseases and a number of other factors. All these processes are characterised by the following common mathematical properties: non-linear dependencies, lag effects, a large number of variables with complex interactions, and the presence of stochastic quantities.

A characteristic feature of ecological systems is that their behaviour depends on a large number of interrelated factors, which are difficult to fully take into account. In this connection, various simplifications are made when studying specific problems. Various types of equations are used to describe simplified models: ordinary differential equations, integro-differential and partial differential equations, lagged equations, stochastic equations, etc.; all of them are derived by a uniform method, for which the law of population (species) conservation is used.

In this paper we study a model of a chemostat with two populations of microorganisms described by a system of Michaelis-Menten equations. Numerical modelling of periodic processes in the chemostat with a harmonic control function was performed using MatLab 6. The solution of the system was constructed by the built-in function ode45 using the Runge-Kutta algorithm (4.5 order of accuracy) for solving non-rigid systems of differential equations. The results were used to plot population and substrate concentrations at different time intervals of the modelling process. The modelling time was chosen sufficient for the end of transient processes and establishment of the regime of periodic oscillations in the chemostat (steady-state regime).

Based on the results of numerical modelling, conclusions were drawn about the effectiveness of periodic modes of chemostat control, allowing to ensure the levels of useful biomass concentrations, which are unattainable under constant control actions.

CHAPTER 1

MICROBIAL COMMUNITIES

Mixed crops are crops in which fermentation of the substrate always consists by means of a mixture of two or more organisms.

Fermentation using mixed cultures offers a number of advantages over traditional single-crop fermentation:

The yield of the finished product may be higher. Yoghurt is produced by fermenting milk with thermophilic Streptococcus and Bulgarian bacillus. It was shown that when these species were grown separately, 24 mmol and 20 mmol acids were obtained, respectively; together with the same amount of inoculum, a yield of 74 mmol was obtained when a community of these cultures was grown. The number of S. thermophilus cells increased from 500 6 10^6 cells per millilitre to 880^6 10^6 cells per millilitre when grown with Bulgarian bacillus.

Growth rates may be higher. In a mixed culture of microorganisms, one of the cultures may produce essential growth factors, or compounds necessary for the life of the other culture such as carbon or nitrogen sources. This may alter the pH of the medium, thereby improving the activity of one or more enzymes. Even the temperature can be raised and favour the growth of the second microbe.

Mixed cultures can lead to multi-step transformations that would be impossible for a single microorganism.

Mixed cultures allow better utilisation of substrates. The substrate for fermented food production is always a complex mixture of carbohydrates, proteins and fats. Mixed cultures have a wider range of enzymes and are able to attack a greater variety of compounds. In addition, with proper strain selection, they are better able to modify or destroy toxic or harmful compounds that may be in the fermentation substrate.

Disadvantages

Scientific study of mixed cultures is difficult. It is obviously more difficult to study fermentation if more than one microorganism is involved. This is why most biochemical studies are done in single culture fermentations, because one variable is removed.

One of the most serious problems in mixed culture fermentation is controlling the optimum balance between the participating microorganisms. This can, however, be overcome if the behaviour of the microorganisms is understood and this information is applied to their control.

For example, bacteria and fungi of marine origin are promising sources of new bioactive compounds that are important for drug discovery programmes. However, only a fraction of biosynthetic genes are transcribed under conventional laboratory conditions, which include cultivation of axenic microbial strains. In contrast, co-culture (also called mixed fermentation) of two or more different microorganisms attempts to mimic the natural situation, because microorganisms always coexist in complex microbial communities. The competition or antagonism experienced in co-cultivation results in a significant increase in microbial productivity and/or in the accumulation of cryptic compounds that are not found in axenic cultures of the producing strain.

CHAPTER 2

MATHEMATICAL MODELS

Models that are analytically explorable and have properties that allow describing a whole range of natural phenomena are called basic models. Basic models, as a rule, are studied in detail in various modifications. Once the essence of processes on such a basic model is thoroughly studied mathematically, the phenomena occurring in much more complex real systems become understandable by analogy. Thus, due to their simplicity and clarity, basic models become extremely useful in the study of a wide variety of systems.

All biological systems at different levels of organisation, from biomacromolecules down to populations, are thermodynamically non-equilibrium, open to flows of matter and energy. Therefore, nonlinearity is an inherent property of the basic systems of mathematical biology. Despite the great diversity of living systems, we can identify some of the most important inherent qualitative properties: growth, self-limitation of growth, ability to switch - existence in two or more stationary modes, auto-oscillatory modes (biorhythms), spatial heterogeneity, quasi-stochasticity. All these properties can be demonstrated on relatively simple nonlinear dynamic models, which act as basic models of mathematical biology.

One of the fundamental assumptions underlying all growth models is that the growth rate of a population is proportional to its population size. This assumption is based on the well-known fact that the most important characteristic of living systems is their ability to reproduce. For many unicellular organisms or cells that are part of cellular tissues, it is simply division, i.e. doubling of the number of cells after a certain interval of time, called the characteristic time of division. For complexly organised plants and animals, reproduction follows a more complex law, but in the simplest model we can assume that the rate of reproduction of a species is proportional to the number of that species.

Fibonacci series

The formulation of mathematical problems in terms of population dynamics dates back to ancient times. It is human nature to reason about subjects that are vitally close to him, and what can be closer than the laws of reproduction of populations - people, animals, plants.

The first extant mathematical model of population dynamics is given in the book "Treatise on Counting" *("Liber abaci"),* dated 1202, written by the greatest Italian scientist Leonardo Fibonacci - Leonardo of Pisa (presumably 1170-1240). This book deals with the following problem. "A certain person raises rabbits in a space enclosed on all sides by a high wall. How many pairs of rabbits are born in one year from one pair, if after a month a pair of rabbits produces another pair, and rabbits give birth starting from the second month after their birth". The solution to the problem is a series of numbers:

$$1, 1, 2, 3, 5, 8, 13, 21, 34, 55, 89, 144, 233, 377...$$

The first two numbers correspond to the first and second month of breeding. The next 12 numbers correspond to the monthly growth of the rabbit population. Each subsequent row is equal to the sum of the two previous ones. The series (1.1) has gone down in history as the Fibonacci series, and its terms are Fibonacci numbers. It is the first recursive sequence of numbers known in Europe (in which the relationship between two or more members of a series can be expressed as a formula). The recurrence formula for the terms of the Fibonacci series was written down by the French mathematician Albert Girer in 1634.

$$U_{n-2} \quad U_{n-1} \quad + \quad U_n$$

Here U represents a member of the sequence, and the lower index is its number in the series of numbers. In 1753, Robert Simpson, a mathematician from Glasgow, noticed that as the ordinal number of the members of a series increases, the ratio of the subsequent member to the previous one approaches the number a, called the Golden Ratio, equal to 1.6180.... Since then, naturalists have observed its regularities in the arrangement of scales on cones, petals in a sunflower, in the spiral formations of mollusk shells and other creations of nature. Fibonacci series and its properties are also used in computational mathematics to create special counting algorithms.

Unlimited growth. Exponential growth.

The second world-famous mathematical model, which is based on the problem of population dynamics, is the classical model of unlimited growth - geometric progression in discrete representation,

$$An + 1 = qAn \tag{1.1}$$

or exponent in continuous

$$\frac{dx}{dt} = Rx \tag{1.2}$$

Here R in general can be a function of both the population itself and time, or it can depend on other external and internal factors.

The assumption that the rate of population growth is proportional to its population size was made as early as the 18th century by Thomas Robert Malthus (1766-1834) in his book On Population Growth (1798). According to the law (1.4), if the proportionality coefficient $R=const$ (as it was assumed by Malthus), the population will grow indefinitely exponentially.

$$x = x_0 e^{rt} \tag{1.3}$$

For most populations, there are limiting factors and for one reason or another, population growth stops. No population in nature grows indefinitely. Hence, there are reasons that prevent such growth. Therefore, the law of exponential growth is valid at a certain stage of growth for populations of cells in tissue, algae or bacteria in culture.

Restricted growth. Verhulst equation.

The basic model describing limited growth is the Verhulst model

$$\frac{dx}{dt} = rx(1 - \frac{x}{K}) \qquad (1.4)$$

This logistic equation has two important properties. For small x, the population size x increases exponentially (as in Equation 1.4), and for large x, it approaches a certain limit K. This value, called population capacity, is determined by the limited food resources, nesting sites, and many other factors that may be different for different species. Thus, ecological niche capacity is a systemic factor that determines the limitation of population growth in a given habitat.

The graph of the dependence of the right part of equation (1.4) on the number x and population size on time are presented in Fig. 1 (a and b).

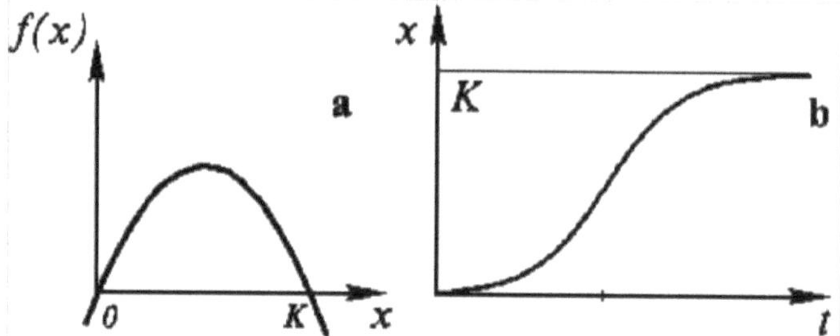

Figure 1. Limited growth. Dependence of the growth rate on population (a) and population on time (b) for the logistic equation (1.4).

In nature, populations have not only a maximum population size, determined by the size of the ecological niche K, but also a minimum critical population size. If the population falls below this critical number due to unfavourable conditions or as a result of predation, it becomes impossible to restore the population.

The lower critical density varies from species to species. Studies by biologists have shown that it can be as low as a couple of individuals for the muskrat and hundreds of thousands of individuals for the American wandering pigeon per thousand square kilometres. It was difficult to assume in advance that such an abundant species had already crossed the critical limit of its numbers and was doomed to extinction. For example, for blue whales, the critical population limit has been found to be in the tens to hundreds.

Mono and Michaelis-Menten models.

One of the reasons for growth limitation may be lack of food - substrate limitation (in the language of microbiology). Microbiologists have long noticed that under conditions of substrate limitation, the growth rate increases in proportion to the substrate concentration, and if there is enough substrate, it reaches a constant value determined by the genetic capabilities of the population. For some time the population grows exponentially until the growth rate starts to be limited by some other factors. This means that the dependence of the growth rate R in formula (1.4) on the substrate can be described in the form:

$$\mu(s) = \frac{\mu S}{K_M + S}$$

Here K_M is a constant equal to the substrate concentration at which the growth rate is half of the maximum; μ is the maximum growth rate equal to the value of r in the formula. This equation was first written by the major French biochemist Jacques Monod (1912-1976).

The Monod model in form coincides with the Michaelis-Menten equation (1913), which describes the dependence of the rate of an enzymatic reaction on the substrate concentration, provided that the total number of enzyme molecules is constant and much less than the number of substrate molecules:

$$\mu(s) = \frac{\mu S}{K_M + S}$$

Here K_M is the Michaelis constant, one of the most important values for enzymatic reactions, determined experimentally, having the meaning and dimension of the substrate concentration at which the reaction rate is half of the maximum.

Models of interaction between two populations. Classical Lotka and Volterra models.

The first in-depth mathematical study of the regularities of the dynamics of interacting populations was given in V. Volterra's book "Mathematical Theory of the Struggle for Existence" (1931) The major Italian mathematician Vito Volterra, the founder of mathematical biology, proposed to describe the interaction of species in the same way as it is done in statistical physics and chemical kinetics; as multiplicative terms in the equations (products of the numbers of interacting species).

The systems studied by Volterra consist of several biological species and a food stock that is utilised by some of the species under consideration. The following assumptions are formulated about the components of the system:

1. Food is either available in unlimited quantities or its supply is strictly regulated over time. Individuals of each species die off so that a constant proportion of existing individuals die per unit of time.

2. Predatory species eat their prey, and the number of prey eaten per unit of time is always proportional to the probability of meeting individuals of the two species, i.e. the product of the number of predators by the number of prey.

3. If there is food in unlimited quantity and several species capable of consuming it, the proportion of food consumed by each species per unit time is proportional to the number of individuals of that species, taken with some species-dependent coefficient (interspecific competition models).

4. If a species feeds on food that is available in unlimited quantities, the increase in species abundance per unit time is proportional to the abundance of the species.

5. If a species feeds on food available in limited quantities, its reproduction is governed by the rate of food intake, i.e. per unit time, growth is proportional to the amount of food eaten.

The above hypotheses allow describing complex living systems by means of systems of ordinary differential equations, in the right parts of which there are sums of linear and bilinear terms. Such equations also describe systems of chemical reactions.

In general, taking into account the self-limitation of abundance by the logistic law, the system of differential equations describing the interaction between the two species can be written in the form:

$$\begin{cases} \dfrac{dx_1}{dt} = a_1 x_1 + b_{12} x_1 x_2 \\[2mm] \dfrac{dx_2}{dt} = a_2 x_2 + b_{21} x_1 x_2 \end{cases}$$

Here, parameters **ai are the** intrinsic growth rate constants of species, **bij are the** species interaction constants, **$(i,j=1,2)$**. The correspondence of the signs of these last coefficients to different types of interactions is given in Table 1.

Table 1: Types of species interactions

Тип взаимодействия	b12	b22	Коэфициенты
Симбиоз	+	+	b12, b21>0
Комменсализм	+	0	b12>0, b21=0
Хищник-жертва	+	-	b12,>0, b21<0
Аменсализм	0	-	b12,=0, b21<0
Конкуренция	-	-	b12, b21<0
Нейтрализм	0	0	b12, b21=0

Examining the properties of type models leads to some important conclusions regarding the outcome of species interactions.

The competition equations (b12>0, b21<0) predict the survival of one of the two species if the intrinsic growth rate of the other species is less than some critical value.

Symbiosis is a long-term close cohabitation of two organisms of different species in which they mutually benefit each other.

Commensalism - copulation, sponging, cohabitation of animals of different species, characterised by the fact that one of them (commensal) permanently or temporarily lives at the expense of the other without causing harm to it.

Amensalism - (from Greek a - negative particle and Latin mensa - table, meal), a form of relationship between organisms, useful for one species, but harmful for another.

Phase portraits

If the system equations are represented in normal form, the state vector of the system uniquely defines its state. Each state of the system in the state space corresponds to a point. The point corresponding to the current state of the system is called the image point. When the state changes, the image point describes a trajectory. This trajectory is called a phase trajectory. The set of phase trajectories corresponding to different possible initial conditions is called a phase portrait.

The phase trajectory and phase portrait can be visualised in the case of a two-dimensional phase space. A two-dimensional phase space is called a phase plane.

Phase plane is a coordinate plane in which two variables (phase coordinates) are plotted along the coordinate axes, which uniquely determine the state of the second-order system. The method of analysing and synthesising a control system based on the construction of a phase portrait is called the phase plane method.

Only one trajectory can pass through any point of phase space. However, there are special points on the phase plane - points where the phase velocity is zero, and hence these points are the equilibrium position of the system. More than one phase trajectory can pass through special points.

To construct the phase portrait, the isocline method is used - lines are drawn on the phase plane that intersect the integral curves at one particular angle. The isocline equation is easily obtained from (1.8)

$$\frac{dx}{dy} = A,$$

$$(1.8)$$

where A is a certain constant value. The value of A represents the tangent of the angle of inclination of the tangent to the phase trajectory and can take values from $-\infty$ to $+\infty$. Substituting A instead of dy/dx in (1.8), we obtain the isocline equation:

$$A = \frac{Q(x,y)}{P(x,y)}$$

$$(1.9)$$

Equation (1.9) defines at every point in the plane a single tangent to the corresponding integral curve except at the point where P $(x,y) = 0$, Q $(x,y) = 0$, where the direction of the tangent becomes undefined because the value of the derivative becomes undefined:

$$\left.\frac{dy}{dx}\right|_{x=\bar{x}, y=\bar{y}} = \frac{Q(\bar{x}, \bar{y})}{P(\bar{x}, \bar{y})} = \frac{0}{0}$$

(1.10)

This point is the point of intersection of all isoclines - a special point. In this point the time derivatives of the variables x and y simultaneously go to zero.

$$\left.\frac{dx}{dt}\right|_{\bar{x},\bar{y}} = P(\bar{x}, \bar{y}) = 0, \quad \left.\frac{dy}{dt}\right|_{\bar{x},\bar{y}} = Q(\bar{x}, \bar{y}) = 0.$$

(1.11)

Thus, at the special point the rates of change of the variables are equal to zero. Thus, the special point of the differential equations of phase trajectories (1.11) corresponds to the stationary state of the system (1.10), and its coordinates are the stationary values of the variables x, y.

Of particular interest are the main isoclines:

$dy/dx=0$, $P(x,y)=0$ - isoclines of horizontal tangents and

$dy/dx=\infty$, $Q(x,y)=0$ are isoclines of vertical tangents.

Having constructed the main isoclines and found the point of their intersection (x, y), the coordinates of which satisfy the conditions:

$$P(\bar{x}, \bar{y}) = 0, \quad Q(\bar{x}, \bar{y}) = 0,$$

(1.12)

We will thus find the point of intersection of all isoclines of the phase plane, where the direction concerning the phase trajectories is uncertain. This is a special point corresponding to the stationary state of the system (Fig. 1.2).

The system (1.10) has as many stationary states as there are intersection points of the main isoclines on the phase plane.

Each phase trajectory corresponds to a set of motions of a dynamic system, pass through the same states and differ from each other only by the beginning of time reference.

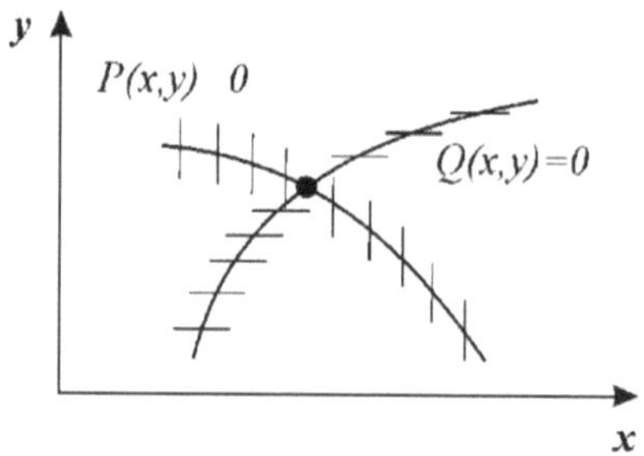

Figure 1.2 - Intersection of the main isoclines on the phase plane.
Thus, the phase trajectories of the system are projections of integral curves in the space of all three dimensions x, y, t onto the dam x, y (Fig. 1.3).

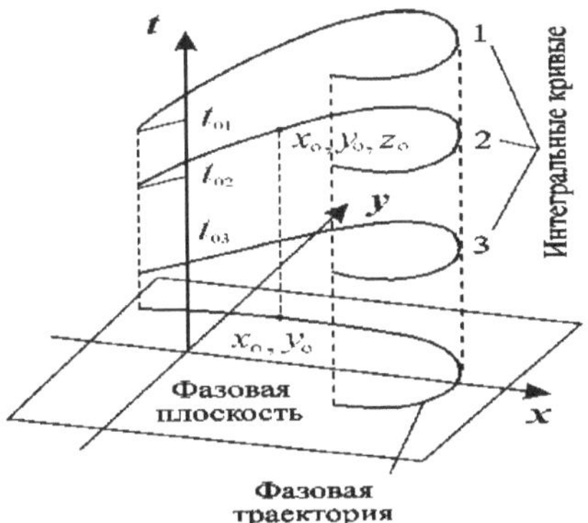

Figure 1.3 - Trajectory of the system in *(x, y, t)* space.

If the conditions of Cauchy's theorem are fulfilled, then a single integral curve passes through each point of space x, y, t. The same is true, due to autonomy, for phase trajectories: a single phase trajectory passes through each point of the phase plane.

Historically, mathematical microbiology as an independent science traces its origins back to Monod's seminal work. Studying the growth of microorganism culture on limiting substrates, Monod showed that when external conditions are constant, the ratio remains constant:

$$Y = (X - X_0)|(S_0 - S)^{-1} = \text{const} \qquad (1.12)$$

Or, in modern terminology, the economic coefficient; the current and initial concentrations of biomass and substrate, respectively. Further, Monod in the same work wrote down a closed system of equations describing the growth of a crop sown in some vessel:

$$dX/dt = \mu\,(S)\,X = \mu_{max}\,S/(K_S + S),$$
$$dX/dt + Y\,dS/dt = 0, \quad X\,(0) = X_0, \quad S\,(0) = S_0, \qquad (1.13)$$

and found an analytical solution. In dimensionless variables:

$$x = X/K_S Y, \quad y = S/K_S, \quad \tau = t\mu_{max} \qquad (1.14)$$

The relationship is expressed in the sum of logarithms:

$$\tau\,(x) = \frac{1 + x_0 + y_0}{x_0 + y_0}\,\ln\frac{x}{x_0} + \frac{1}{x_0 + y_0}\,\ln\frac{y_0}{x_0 + y_0 - x} \qquad (1.15)$$

(Figure 1.4, curve I).

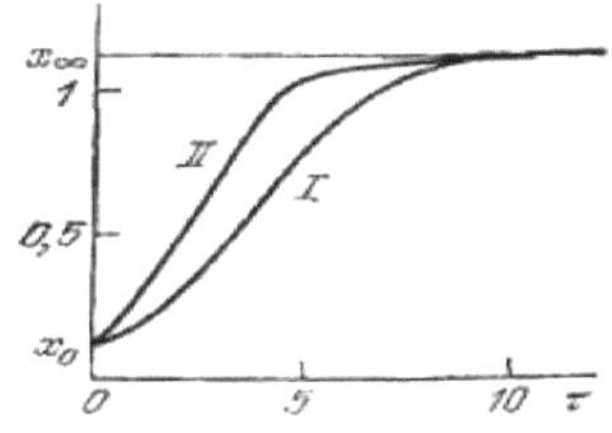

Figure 1.4 - Biomass growth: 1 Mono model (1.15), 2 Verhulst model

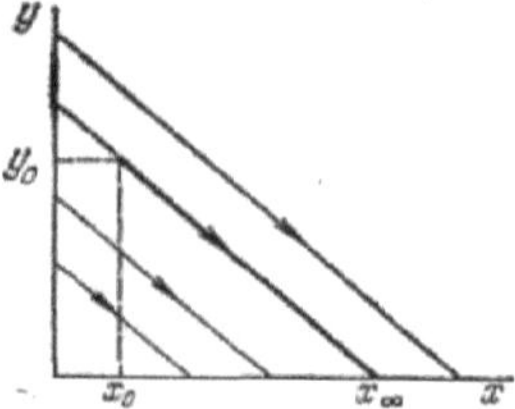

Figure 1.5 - Phase portrait of the system (1.15).

Thus, already in the 40s, a mathematical model of bacterial culture growth in a closed vessel was constructed. The cessation of biomass growth in this model has a clear physical cause - depletion of nutrient medium, and this, in our opinion, is the advantage of the Mono model (1.15) over others, for example, over the Verhulst model.

Model II. The next stage in mathematical microbiology can be considered the development of models of microbial growth under flow-through conditions. Flow-through, or continuous cultivation has long had a wide application in industrial production, as it provides a continuous output of a homogeneous mass of cells through the plant. The theory of the continuous cultivator, the Chemostat, was proposed in the 50's by Mono and in the USA in parallel by Novick and Scillard .

The equations of biomass and substrate limiting balance in the chemostat (using Mono's formula were written as a system of equations

$$\frac{dX}{dt} = \mu(S)X - DX, \qquad \mu(S) = \frac{\mu_{max} S}{K_S + S},$$

$$\frac{dS}{dt} = -\frac{1}{Y}\mu(S)X + D(S_0 - S),$$

$$(1.16)$$

where is the flow rate, or dilution rate, having the dimension of substrate concentration entering the cultivator, the economic coefficient. In dimensionless variables we have:

$$dx/d\tau = xy/(1+y) - \delta x, \quad dy/d\tau = -xy/(1+y) + \delta(y_0 - y). \qquad (1.17)$$

This is where the parameters are entered

$$\delta = D/\mu_{max}, \quad y_0 = S_0/K_S.$$

The system has two special points:

$$\bar{x} = 0, \quad \bar{y} = y_0,$$

$$\bar{x} = y_0 - \bar{y}, \quad \bar{y} = \delta \cdot (1 - \delta)^{-1}.$$

$$(1.18)$$

The limiting value of the flow rate is called the leaching rate. Indeed, when biomass growth can no longer compensate for its outflow and the crop is completely "washed out" from the cultivator. Stationary concentrations x and in (1.17) are represented in Fig. 1. 6 by solid lines.

Fig. 1.7 shows the phase portrait of the system (1.17). In case (a) all trajectories converge to stationary point 2, otherwise to point 1 (biomass leaching occurs).

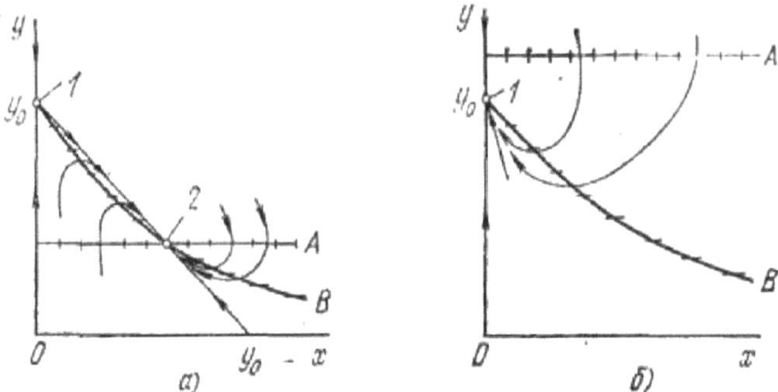

Figure 1.6 - Dependence of stationary concentrations on 8, solid lines - model II, dotted - model I

Figure 1.7 - Phase portrait of model II (Mono). A - isoclines of vertical tangents: B - horizontal tangents

Thus, in a flow-through cultivator, a steady state equilibrium with non-zero biomass can exist when microbial growth is limited by substrate. The stability of the system is ensured as if by a "negative feedback" due to the concentration of the limiting substrate.

An important property of chemostat - "autostabilisation" of the limiting factor - was noted in the work of Pechurkin et al.

Let us note here another property of the chemostat, first investigated by Novik and Scillard autoselection of mutants: if as a result of mutation a new strain, a better utilitarian of a given substrate, is formed, then over time it displaces the output strain from the chemostat. In other words, two strains competing for the same substrate cannot sustainably coexist in a flow-through cultivator.

$$dx/d\tau = xy/(1 + y + \gamma y^2) - \delta x,$$
$$dy/d\tau = -xy/(1 + y + \gamma y^2) + \delta (y_0 - y). \tag{1.19}$$

Model III. Let us now turn to the model that takes into account substrate suppression (reduction of growth rate at high substrate concentrations). In this case, the dimensionless growth rate as a function of dimensionless substrate concentration is given by expression (1.11). The function reaches its maximum value at (see Figure 1.7). System of equations describing biomass growth in a chemostat in the presence of substrate suppression (1.19)

Thus, if there is only one positive stationary value of y and the behaviour of the system does not practically differ from that of the Mono model (in this case we do not go to the maximum of the curve and the influence of substrate suppression is not affected).

From formula (1.19) it follows that the point corresponding to the lower part of the curve is a stable node, and the upper one is a saddle. In the interval of parameter values, there are two stable special points simultaneously and, therefore, hysteresis will be observed at a smooth change of the parameter. (Fig. 1.8).

The phase portrait of the system (1.19) in the case when all three special points exist simultaneously is shown in Fig. 1.9. The isoclines of horizontal tangents intersect the stationary value line three times. On the phase plane there exists a separatrix separating the zones of attraction of stable points 1 and 3.

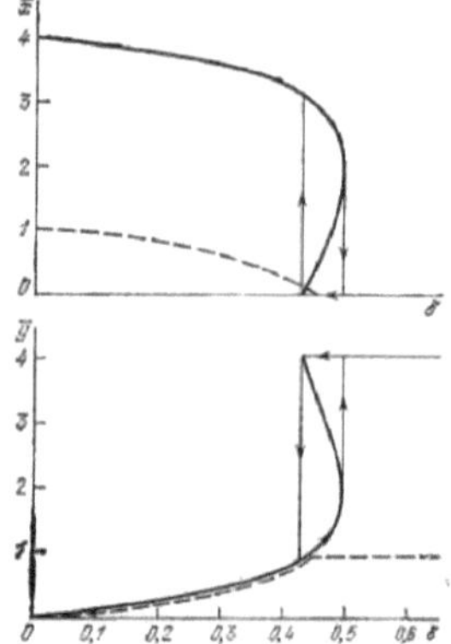

Figure 1.8 - Dependence of stationary concentrations x and y on *6* (model III)

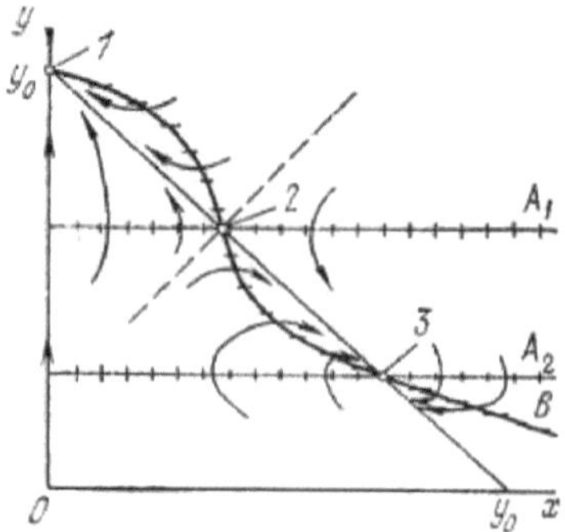

Figure 1.9 - Phase portrait of model III. Isoclines of vertical tangents to horizontal tangents - curve B - dotted-separatrix of the saddle point 2.

CHAPTER 3

MATHEMATICAL MODELS OF FLOW-THROUGH CULTURE OF MICROORGANISMS

An empirical approach to model building is generally accepted in microbiology. From all factors affecting cell growth, a limiting factor is selected and the dependence of the population growth rate on substrate concentration is found experimentally. In general, the kinetics of cell concentration in a continuous culture is described by the equation:

$$\frac{dx}{dt} = x(\mu - v)$$

Here x is the concentration of cells in the cultivator; μ- is a function describing population reproduction. It may depend on cell concentration x, substrate concentration (usually denoted by S), temperature, pH of the medium and other factors; V - washout rate.

External regulators are required to maintain the crop in the area of unlimited growth. If growth is limited by an external factor, such as a lack of substrate, the steady-state mode of operation of the cultivator is established by self-regulation. This occurs in natural flow systems and in the most common type of continuous cultivator, the chemostat, where the culture dilution rate or flow rate is set. Chemostat theory was first developed by Monod (1950) and Herbert (1956) and has been continually refined since then. Modern models take into account structural heterogeneity of biomass, age heterogeneity of culture (distribution of numbers of individuals by age in the population), discrete and continuous representation of age structure and other details of cultivation.

With continuous stirring, the entire volume of the cultivator can be considered homogeneously filled, the concentrations of substrate and cells at each point of the cultivator are the same, and the behaviour of these concentrations over time can be described using a system of ordinary differential equations:

$$\begin{cases} (a)\dfrac{dx}{dt} = \mu(S)x - Dx, \\[2mm] (b)\dfrac{dS}{dt} = DS_0 - \alpha\mu(S)x - DS, \\[2mm] (c)\mu(S) = \dfrac{\mu_m S}{K_m + S} \end{cases}$$

Here S - substrate concentration; x - concentration of cells in the cultivator; So - concentration of substrate entering the cultivator; D - flow rate (dilution) of the culture; α^{-1} - economic coefficient showing what part of the absorbed substrate goes to biomass increment. The meaning of the terms included in the right-hand sides of the equations:

μ (S)x - biomass growth due to substrate uptake;

17

- Dx - biomass outflow from the cultivator;
- *$\alpha\mu(S)x$ is the* amount of substrate absorbed by the culture cells;
-   S_0 - substrate inflow to the cultivator;
- DS - outflow of unused substrate from the cultivator;

K_m - Michaelis-Menten constant [mol/L].

The biomass growth rate is assumed to depend only on substrate concentration according to Mono's formula (1.12).

In a chemostat, the dilution rate D is set arbitrarily, and microorganisms "choose" their concentration and maintain a specific growth rate μ equal to D. This process of self-regulation of microbial reproduction rate is related to the dependence of specific growth rate μ on the concentration of organic substrate in the environment.

A detailed discussion of this issue was carried out by Monod and Herbert. When the specific growth rate μ is less than a given specific dilution rate D, the population density of microbial cells in the cultivator starts to decrease, i.e. microorganisms are gradually washed out by the medium flow. As a result, the substrate concentration in the cultivator increases (as the rate of substrate utilisation by the microbial population decreases).

As the substrate concentration increases, the specific growth rate increases. When the value of μ becomes equal to D, equilibrium is established in the system. When the specific dilution rate D decreases, the picture is reversed. The inequality $D<\mu$ means that the number of cells in the chemostat increases, while the substrate concentration decreases.

Consequently, the specific growth rate // will also decrease until it becomes equal to D. As a result, a new equilibrium position will be established. It follows from this that continuous processes have the ability to self-regulate. This reasoning is true when $D<\mu_{max}$. otherwise, i.e. when $D\geq\mu_{max}$ max , microorganisms are always washed out of the cultivator.

Cultivators of this type can be connected in series, resulting in multi-stage continuous cultivation, called single-stream multi-stage cultivation by Herbert [2].

Finally, the systems described above can be supplemented by recirculating a certain amount of cell biomass. In this case, the cell suspension leaving the last fermenter is concentrated in a continuous-flow centrifuge. A part of the cells is recirculated (returned) to the first fermenter.

The general term "ecostat" has been proposed for devices in which both abiotic factors (concentrations) and biotic conditions of individuals (in particular density) are regulated.

A mathematical model of the Michaelis-Men gen chemostat.
The system of differential equations describing the process of cultivation of

microbial community in chemostat in the presence of one substrate is written in the form:

$$\begin{cases} \dot{S} = D(S_0 - S) - \sum_{j=1}^{n} \frac{1}{y_j} \mu_j(S)x_j \\ \dot{x}_i = (-D + \mu_j(S))x_i \\ i = 1.....n \end{cases}$$

The substrate is fed with nutrient solution at concentration So and washed out of the cultivator at concentration S, xi is the concentration of microorganisms of i-th population in the chemostat, yi is the growth constant of microorganisms]-th population.

Most of the known dynamic models for chemostat differ from each other in the type of function // (S) of the specific growth rate.

During continuous fermentation in a chemostat, as a rule, all cultivation parameters are maintained at a constant level. However, due to some reasons (physicochemical or biological), some parameters may change, which will naturally lead to a disturbance of the equilibrium state of the fermentation process.

Many authors have experimentally investigated the question of the nature of the behaviour of a cultured population of microorganisms at a sudden change in one or more parameters, such as specific dilution rate D, substrate concentration S, and temperature. As a result of the research it is shown that after a sudden change of a parameter a transient process occurs, as a result of which either the system passes to a new state or the fermentation process stops altogether. In case of a sharp change in the size of the cultivated population, the system, as a rule, returns to the initial stationary state (a sharp decrease in the concentration of living organisms can be caused, for example, by a strong dose of radioactive radiation or some other factor of influence).

If n different species of microorganisms are cultivated on the same organic substrate in a chemostat, there will be a displacement of some species by others. The most detailed analysis (in the experimental and theoretical sense) of microorganism selection processes in a chemostat was first carried out in the works of Moser, Novik and Scilard.

Basic mathematical results for the Mechaelis-Menten chemostat model.

We will consider a dynamic model of Michaelis-Menten chemostat [4] with one organic substrate and two species of microorganisms, in which the specific growth rate of the culture:

$$\mu(s) = \frac{mS}{a + S}$$

The Michaelis-Menten chemostat model is described by the following system of differential equations :

$$\begin{cases} \dot{S}(t) = (u(t) - S(t))D - \dfrac{m_1 x_1(t)S(t)}{a_1 + S(t)} - \dfrac{m_2 x_2(t)S(t)}{a_2 + S(t)} \\[2ex] \dot{x}_1(t) = \left(\dfrac{m_1 x_1(t)S(t)}{a_1 + S(t)} - D \right) x_1(t) \\[2ex] \dot{x}_2(t) = \left(\dfrac{m_2 x_2(t)S(t)}{a_2 + S(t)} - D \right) x_2(t) \end{cases}$$

S(t), xl(t), x2(t) denote the densities of nutrient substrate and microorganisms at time t.

The scalar control function u(t) >= 0 determines the feed rate of the nutrient substrate to the chemostat.

The remaining parameters D, ml, al, m2, a2 of the model (1.16) are given positive numbers.

The number D (leaching coefficient) determines the rate of substance flow through the chemostat, the numbers mi [mol/L/sec] (i=1,2) determine the maximum growth rate of the i-th population, and the numbers ai [mol/L]- (Michaelis-Menten constants) denote the substrate density at which the specific growth rate - miS/(ai + S) for the i-th population is equal to half of the maximum value of mi

It should be noted that, strictly speaking, the results are valid only for small values of the amplitudes of the input influences um, since in the analysis of oscillatory processes an approximate method is used in which the solutions S(t), xl(t) and x2(t) of the system are sought in the form of asymptotic expansions by powers of the value um, which plays the role of a small parameter.

CHAPTER 4

NUMERICAL ANALYSIS OF PERIODIC PROCESSES IN CHEMOSTAT

A system of differential equations describing the processes of development of microbial populations in a chemostat proposed by Michaelis-Menten is given:

$$\begin{cases} \dot{S}(t) = (u(t) - S(t))D - \dfrac{m_1 x_1(t)S(t)}{a_1 + S(t)} - \dfrac{m_2 x_2(t)S(t)}{a_2 + S(t)} \\[2mm] \dot{x}_1(t) = \left(\dfrac{m_1 x_1(t)S(t)}{a_1 + S(t)} - D \right) x_1(t) \\[2mm] \dot{x}_2(t) = \left(\dfrac{m_2 x_2(t)S(t)}{a_2 + S(t)} - D \right) x_2(t) \\[2mm] u(t) = u_0 + u_m \sin(\omega t) \end{cases}$$

The software used to build the system solutions was MathWorks Inc. - MatLab 6.12R. The MATLAB system (MATrix LABoratory) has long been successfully developed by MathWorks. Matlab is a high-performance language for technical calculations and interactive system in which the main data element is an array. It allows solving various tasks related to calculations, especially those involving matrices and vectors, several times faster than when writing programmes using C or Fortran programming languages.

The choice of this programme was justified by the fact that MatLab is oriented to numerical methods and has several methods for solving systems of differential equations (SDEs) in Cauchy form, including those for stiff systems.

In MATLAB system there is a whole package of procedures designed to solve the Cauchy problem for a system of ordinary differential equations (ODE). Moreover, this package provides different algorithms for solving such a problem, but in order to save effort and facilitate work, the handling and writing of the required additional code is the same and does not depend on the algorithm. Let us consider the method of solving such systems, regardless of the solution algorithm used in Matlab 6.

A large class of ODEs, namely, equations with a single independent variable, most often referred to as time t, solved with respect to the senior derivative, can be reduced to a system of first order differential equations of the form:

$$y'(t) \quad F(t, y(t))$$

If the right-hand sides of the corresponding system are smooth enough, then the described system has a single solution. And if, and in addition, there is no analytical expression for the solution, then the solution can be obtained numerically using one of the algorithms used in MATLAB.

Methods for solving ODEs (Solvers).

In MatLab, different procedures can be used to numerically solve ODEs.

21

MATLAB has the following functions for solving non-rigid systems of equations: ode45 - based on the explicit Runge-Kutta method. It is a one-step algorithm - to calculate y(tn) it is necessary to know the solution at one previous pointu(1p-1). This function is most convenient for the first, 'shooting' solution of most problems. ode23 - is also based on the explicit Runge-Kutta method, but of lower order, so it is more suitable for obtaining a coarser solution (with lower accuracy) and in the presence of small rigidity. It is also a one-step method.

ode113 - uses the variable order Adams-Bashforth-Milton method. It may be more efficient than the ode45 method, especially at high accuracy and when the right-hand sides of the equations are difficult to calculate. The method is multistep, so it requires knowledge of the solution at several initial points to start solving.

MATLAB provides 4 functions for solving rigid systems of equations.

ode 15s is based on a numerical backward differentiation method known as the Tyre method. Also, like the odell3 method, this method is multi-step. If you have a tough problem or could not solve it with ode45, try ode 15s.

ode23s uses the second-order Rosenbrock method. Since it is a one-step method, it may be more efficient than ode15 s method for low accuracy cases. The ode23t is an implementation of the free multiplier trapezoid rule. This method makes sense to use only if the problem is moderately stiff and there is no need for numerical damping of the solution.

ode23tb implements a two-stage solution to the implicit Runge-Kutta formula. Like the method

ode23s this method is effective when the required accuracy of the solution is low.

General rules for calling ODU solvers

All of the above functions are called in the same way. In the simplest form this looks as follows:

$$[T,Y] = odeXX(odefun,tspan,y0,options,p1,p2,...).$$

Where:

odeXX - any of the functions listed above;

odefun - string containing the name of the function describing the right parts of the system;

tspan is a vector defining the integration interval. If the vector tspan=[tO tfinal] has only two elements, the integration goes from tO to!final. If the vector tspan has more than two elements, the function odeXX outputs

solution at all points that are listed in the vector tspan, tO > tfinal is admissible;

y0 - vector of initial conditions of the problem;

T -vector -column of time moments;

Y is a matrix of solutions. Each row of the matrix contains a vector of solutions at time ti y(i))

for the corresponding moment in time.

The options argument is set in a special way using the options set function (see the odeset Function help).

* 1,p2, - additional parameters are passed to the function of definition of right parts in the form:

* defun(t,y,flag,p1,p2,...) - the function of defining the right parts of ODEs for the system has the form:

function dydt = cheMehMenten(t,y,U0, Um, w, D, a1, a2, m1, m2)

U=U0+Um*sin(w*t);

dydt = [

(U-y(1))*D-(m1*y(1)*y(2))/(a1+y(1))-(m2*y(3)*y(1))/(a2+y(1))

((m1*y(1))/(a1+y(1))-D)*y(2)

((m2*y(1))/(a2+y(1))-D)*y(3)

];

A computational experiment.

Computational experiment is a research technology based on the construction and analysis of mathematical models (MM) of the phenomena and objects under study by means of computers. The main procedures of computational experiment are:

* building a mathematical model of a process (object) in the space of continuous independent variables
* selection (construction) of a numerical method for solving the MM equations, i.e. formation of a discrete MM
* software implementation of the numerical method
* programme testing and debugging
* calculating
* analysing the results and possibly correcting the model.

Let us note some peculiarities of the listed procedures. Typical mathematical models of physical processes are systems of differential and algebraic equations, usually nonlinear. It is extremely rare to obtain their solutions in analytical form (at best, it is possible to prove only the existence of a solution). In order to obtain quantitative characteristics of the processes, it becomes necessary to involve a computer and, as a consequence, to build a discrete model of the process. The way of forming the latter is determined by the chosen numerical method. In general, a discrete model is a system of algebraic equations approximating the original differential equations and an algorithm for solving these equations. The transition to a discrete model is associated with the replacement of continuous independent variables by their discrete analogues; accordingly, the description of the phenomena and processes under study can be obtained only in the form of grid functions. The solution of the discrete model equations requires the development of an appropriate programme and its testing on similar problems whose solution is known. The final stage of the computational experiment is to perform the necessary calculations, analyse the results from the point of view of their correspondence to the process under study and, if necessary, correct the model. Thus, the computational experiment contains three main elements: model - discrete model - programme.

CHAPTER 5

DESCRIPTION OF TECHNOLOGICAL PROCESSES IN THE BIOREACTOR.

Bioreactor (fermenter) - a container in which microorganisms are grown, the process carried out by microorganisms is called fermentation.

The fermentation process is divided into six main stages:

1.	Creating the environment. First of all, an appropriate culture medium must be selected. Microorganisms need organic carbon sources, a suitable nitrogen source and various minerals for their growth. In the production of alcoholic beverages, the medium should contain sedimented barley, fruit or berry pomace. For example, beer is usually made from malt mash and wine from grape juice. In addition to water and possibly some additives, these extracts make up the growth medium.

The media for making chemicals and drugs are much more complex. Most often sugars and other carbohydrates, but often oils and fats, and sometimes hydrocarbons, are used as carbon sources. Nitrogen sources are usually ammonia and ammonium salts, as well as various products of plant or animal origin: soybean meal, soybean meal, cotton seed meal, peanut meal, corn starch by-products, slaughterhouse waste, fish meal, yeast extract. The formulation and optimisation of growth media is a highly complex process and the recipes for industrial media are a jealously guarded secret.

2.	Sterilisation. The medium must be sterilised to destroy all contaminating microorganisms. The fermenter itself and auxiliary equipment are also sterilised. There are two methods of sterilisation: direct injection of superheated steam and heating by means of a heat exchanger. The degree of sterilisation desired depends on the nature of the fermentation process. It should be maximised in the production of pharmaceuticals and chemicals. The sterility requirements for the production of alcoholic beverages are less stringent. Such fermentation processes are referred to as "protected" because the conditions created in the medium are such that only certain microorganisms can grow in them. For example, in beer production, the growth medium is simply boiled, not sterilised; the fermenter is also used clean, but not sterile.

3.	Obtaining a culture. Before the fermentation process can begin, a pure, highly productive culture must be obtained. Pure cultures of microorganisms are stored in very small volumes under conditions that ensure their viability and productivity; this is usually achieved by storage at low temperature. A fermenter can hold several hundred thousand litres of culture medium and the process is started by introducing a culture (inoculum) of 1-10% of the volume to be fermented. Thus, the initial culture should be grown step by step (with crossings) until a level of microbial biomass sufficient to carry out the microbiological process with the required productivity is reached.

It is absolutely necessary to maintain the purity of the culture during all this time, preventing it from being contaminated by foreign microorganisms. Aseptic conditions can only be maintained by careful microbiological and chemical-technological control.

4.   Growth in an industrial fermenter (bioreactor). Industrial microorganisms must grow in the fermenter under optimal conditions for the formation of the desired product. These conditions are strictly controlled to ensure that they support microbial growth and product synthesis. The design of the fermenter should allow the growth conditions - constant temperature, pH (acidity or alkalinity) and concentration of dissolved oxygen in the medium - to be regulated.

A conventional fermenter is a closed cylindrical tank in which the medium and microorganisms are mechanically mixed. Air, sometimes oxygenated, is pumped through the medium. The temperature is controlled by water or steam flowing through the heat exchanger tubes. This kind of stirred fermenter is used when the fermentative process requires a lot of oxygen. Some products, on the other hand, are formed under oxygen-free conditions, in which case fermenters of a different design are used.

5.   Product isolation and purification. At the end of fermentation, the broth contains microorganisms, unused nutrient components of the medium, various products of microorganisms and the product that was desired to be produced on an industrial scale. This product is therefore purified from the other components of the broth. In the production of alcoholic beverages (wine and beer), it is sufficient to simply separate the yeast by filtration and condition the filtrate. However, the individual chemicals produced by fermentation are extracted from a complex broth. Although industrial microorganisms are specially selected for their genetic properties so that the yield of the desired product of their metabolism is maximised (in a biological sense), the concentration is still small compared to that achieved by chemical synthesis. Therefore, it is necessary to resort to complex extraction methods such as solvent extraction, chromatography and ultrafiltration.

6.   Recycling and elimination of fermentation waste. All industrial microbiological processes produce waste: broth (the liquid remaining after extraction of the product); cells of used microorganisms; dirty water used to wash the plant; water used for cooling; water containing trace amounts of organic solvents, acids and alkalis. Liquid wastes contain many organic compounds; if discharged into rivers, they will stimulate the intensive growth of natural microbial flora, which will deplete river waters of oxygen and create anaerobic conditions. Therefore, the waste is biologically treated before disposal to reduce the organic carbon content.

Industrial microbiological processes.

Industrial microbiological processes can be categorised into 5 main groups:
1)   microbial biomass cultivation;
2)   obtaining products of metabolism of microorganisms;
3)   production of enzymes of microbial origin;

4)   obtaining recombinant products;

5)   Biotransformation of substances.

Cultivation of microbial biomass. Microbial cells themselves can serve as the end product of the production process. Two main types of microorganisms are produced on an industrial scale: yeast, essential for baking, and single-celled microorganisms used as a source of proteins that can be added to human and animal food.

Metabolic products. After introduction of the culture into the nutrient medium, a lag phase is observed, when no visible growth of microorganisms occurs; this period can be considered as a time of adaptation. Then the growth rate gradually increases, reaching a constant, maximum value for the given conditions; this period of maximum growth is called the exponential, or logarithmic, phase. Gradually, growth slows down, and the so-called stationary phase occurs. Then the number of viable cells decreases and growth stops.

Following the kinetics described above, it is possible to follow the formation of metabolites at different stages. In the logarithmic phase, products vital for microbial growth are formed: amino acids, nucleotides, proteins, nucleic acids, carbohydrates, etc. They are called primary metabolites.

Many primary metabolites are of considerable value. For example, glutamic acid (or rather its sodium salt) is a component of many food products; lysine is used as a food additive; phenylalanine is a precursor to the sugar substitute aspartame. Primary metabolites are synthesised by natural microorganisms in quantities necessary only to meet their needs. Therefore, the task of industrial microbiologists is to create mutant forms of microorganisms - superproducers of the corresponding substances. Significant progress has been made in this field: for example, it has been possible to obtain microorganisms that synthesise amino acids up to a concentration of 100 g/l (for comparison, wild-type organisms accumulate amino acids in quantities of milligrams).

In the growth retardation phase and in the stationary phase, some microorganisms synthesise substances that are not formed in the logarithmic phase and do not play an obvious role in metabolism. These substances are called secondary metabolites. They are not synthesised by all microorganisms, but mainly by filamentous bacteria, fungi and spore-forming bacteria. Thus, the producers of primary and secondary metabolites belong to different taxonomic groups. While the question of the physiological role of secondary metabolites in the producer cells has been the subject of serious discussions, their industrial production is of undoubted interest, since these metabolites are biologically active substances: some of them have antimicrobial activity, others are specific enzyme inhibitors, others are growth factors, and many have pharmacological activity. The production of such substances served as a basis for the creation of a number of branches of the microbiological industry. The first in this series was the production of penicillin; the microbiological method of penicillin production was developed in the 1940s and laid the foundation of modern industrial biotechnology.

The pharmaceutical industry has developed highly sophisticated methods for screening (mass testing) microorganisms for their ability to produce valuable secondary metabolites. Initially, the purpose of screening was to produce new antibiotics, but it was soon discovered that microorganisms also synthesise other pharmacologically active substances. During the 1980s, the production of four very important secondary metabolites was established. These were: cyclosporine, an immunosuppressant used as an agent to prevent rejection of implanted organs; imipenem (a modification of carbapenem), the substance with the broadest spectrum of antimicrobial activity of any known antibiotic; lovastatin, a cholesterol-lowering drug; and ivermectin, an antihelminthic agent used in medicine to treat onchocerciasis, or river blindness, and in veterinary medicine.

Enzymes of microbial origin. On an industrial scale, enzymes are obtained from plants, animals and microorganisms. The use of the latter has the advantage of allowing the production of enzymes in huge quantities using standard fermentation techniques. Moreover, it is incomparably easier to increase the productivity of microorganisms than of plants or animals, and the use of recombinant DNA technology makes it possible to synthesise animal enzymes in microbial cells. Enzymes obtained in this way are mainly used in the food industry and related fields. The synthesis of enzymes in cells is genetically controlled, and therefore the available industrial microorganism producers have been obtained by directed modification of the genetics of wild-type microorganisms.

Recombinant products. Recombinant DNA technology, better known as "genetic engineering", allows the genes of higher organisms to be incorporated into the genome of bacteria. As a result, bacteria acquire the ability to synthesise "foreign" (recombinant) products - compounds that previously only higher organisms could synthesise. On this basis, many new biotechnological processes have been developed to produce human or animal proteins that were previously unavailable or used at great risk to health. The term "biotechnology" itself became popular in the 1970s in connection with the development of ways to produce recombinant products. However, the term is much broader and includes any industrial method based on the use of living organisms and biological processes.

The first recombinant protein produced on an industrial scale was human growth hormone. One of the proteins of the blood clotting system, namely factor VIII, is used to treat haemophilia. Before genetically engineered methods were developed to produce this protein, it was isolated from human blood; the use of such a preparation carried the risk of human immunodeficiency virus (HIV) infection.

For a long time, diabetes has been successfully treated with animal insulin. However, scientists believed that a recombinant product would cause fewer immunological problems if it could be obtained in pure form, without impurities of other peptides produced by the pancreas. In addition, it was expected that the number of diabetic patients would increase over time due to factors such as changes in dietary patterns, improved medical care for pregnant women with diabetes and the resulting increased incidence of genetic predisposition to diabetes, and finally, an expected increase in the life expectancy of diabetic patients. The first recombinant insulin went on sale in 1982, and by the late 1980s it had virtually supplanted animal insulin.

Many other proteins are synthesised in the human body in very small amounts, and the only way to produce them on a scale sufficient for use in the clinic is through recombinant DNA technology.

These proteins include interferon and erythropoietin. Erythropoietin, together with myeloid colony-stimulating factor, regulates blood cell formation in humans. Erythropoietin is used to treat anaemia associated with renal failure and may be used as a platelet boosting agent in cancer chemotherapy.

Biotransformation of substances. Microorganisms can be used to transform compounds into structurally similar but more valuable substances. Since microorganisms can only catalyse certain substances, the processes they are involved in are more specific than purely chemical ones. The best known biotransformation process is the production of vinegar by converting ethanol into acetic acid. But among the products formed during biotransformation are such highly valuable compounds as steroid hormones, antibiotics, prostaglandins.

CHAPTER 6

EXPERIMENTAL RESULT

This system of differential equations describing the processes of development of microbial populations in the chemostat proposed by Michaelis-Menten:

$$\begin{cases} \dot{S}(t) = (u(t) - S(t))D - \dfrac{m_1 x_1(t)S(t)}{a_1 + S(t)} - \dfrac{m_2 x_2(t)S(t)}{a_2 + S(t)} \\[2mm] \dot{x}_1(t) = \left(\dfrac{m_1 x_1(t)S(t)}{a_1 + S(t)} - D \right) x_1(t) \\[2mm] \dot{x}_2(t) = \left(\dfrac{m_2 x_2(t)S(t)}{a_2 + S(t)} - D \right) x_2(t) \\[2mm] u(t) = u_0 + u_m \sin(\omega t) \end{cases} \qquad (4.1)$$

Necessary:

1. Using the written programme (Appendix 1), construct a numerical solution of the system (4.1) at time interval t [0; T], where T is chosen sufficient to establish the steady-state regime, with positive parameters of the system recorded in Table 2.

2. Check whether the control method affects the constant population concentration and whether it is possible to achieve a higher population concentration by harmonious influence than with a constant control action.

3. Draw graphs of the solution of the system

Parmeters are set in the programme:

U0 - constant component of harmonic influence;

Um is the amplitude of oscillations;

w is the frequency of oscillation;

So is the initial value of the substrate;

XIo is the initial value of the first population;

X2o is the initial value of the second population;

D, al, a2, ml, m2, - chemostat parameters;

TO is the start time of the simulation;

 Tm - modelling end time

Table 4.1 - Data for simulation modelling of Lactococcus lactis subsp. Lactis and Lactococcus lactis subsp. cremoris cultures.

Параметр	$u()$	um	ω	№ рисунка	Параметры хемостата и начальные условия
	1	0	0	Рис.4.1	$S(0)=1$
	1	0,05	1	Рис.4.2	
Значение	1	0,3	1	Рис.4.3	$X_1(0)=1$
	1	0,5	1	Рис.4.4	$X_2(0)=1$
	1	0,7	1	Рис.4.5	
	1	0,8	1	Рис.4.6	D=1
	1	1	1	Рис.4.7	$a_1=0.25;\ a_2=1;$ $m_1\ 2;\ m_2\ 4,5;$

The chart in the upper left corner shows:

- Maximum values of substrate concentrations (Smax), and populations (X1mah;X2mah) achieved at the interval [T0; Tm];
- Control-type of control function;
- Dt - krok obkuslenie;
- Model - Chemostat parameters;

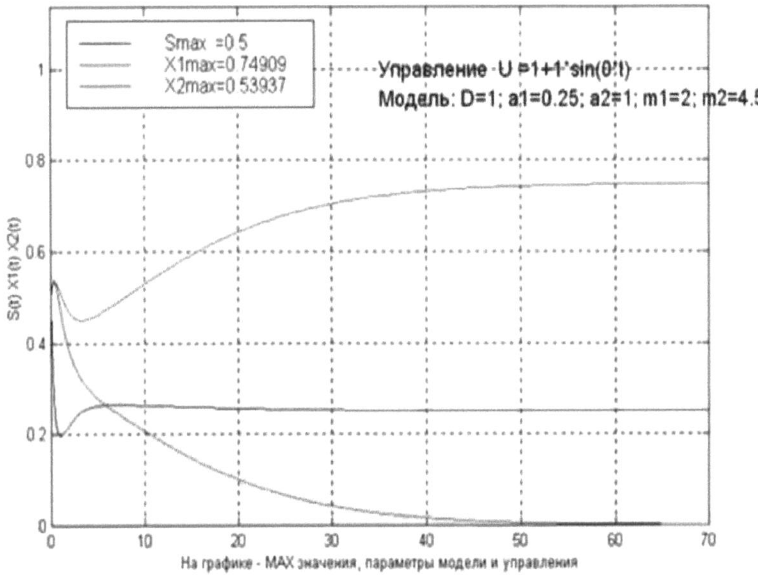

Figure 4.1 - Culture growth character at constant substrate supply. XI - Lactococcus lactis subsp. lactis X2 - Lactococcus lactis subsp. cremoris. S - substrate.

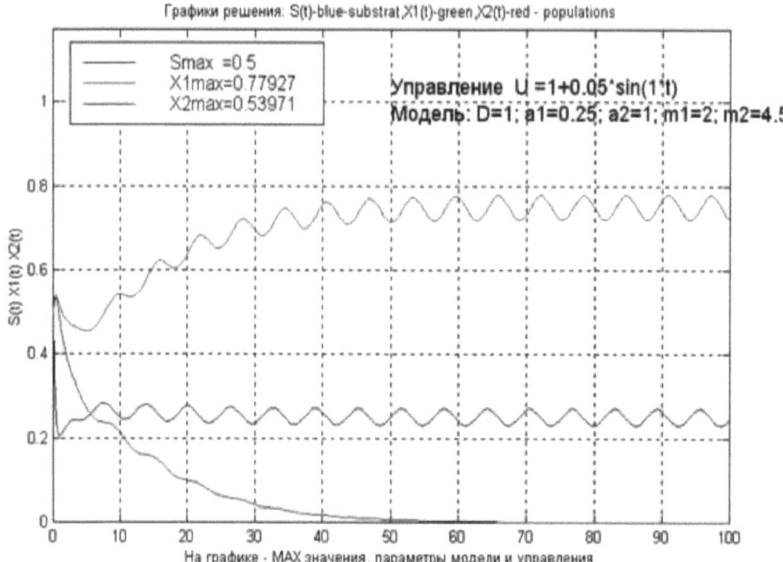

Figure 4.2 - Culture growth character at pulse substrate supply. XI - Lactococcus lactis subsp. lactis X2 - Lactococcus lactis subsp. cremoris. S - substrate. Substrate feeding frequency 0.02 l/xv. Psrud - 10 xv.

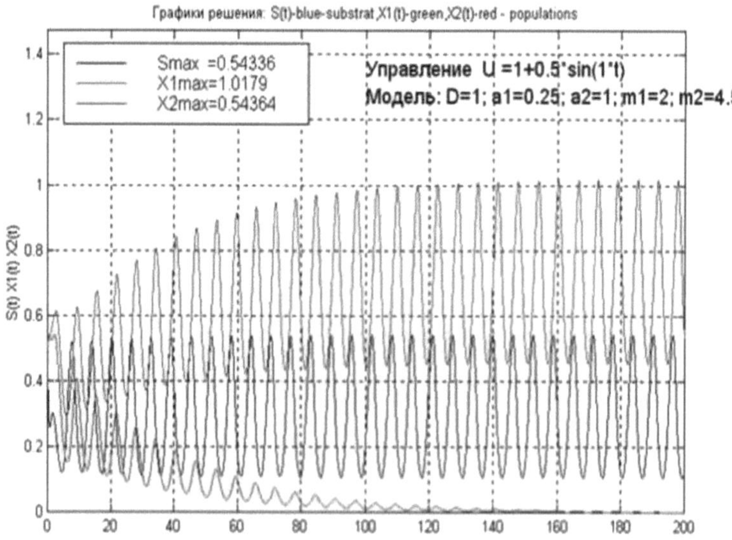

Figure 4.3 - Culture growth pattern during pulsed substrate feeding.
XI - Lactococcus lactis subsp. lactis X2 - Lactococcus lactis subsp. cremoris. S - substrate. Frequency of substrate feeding 0.08 l/xv. Peryud 5 xv

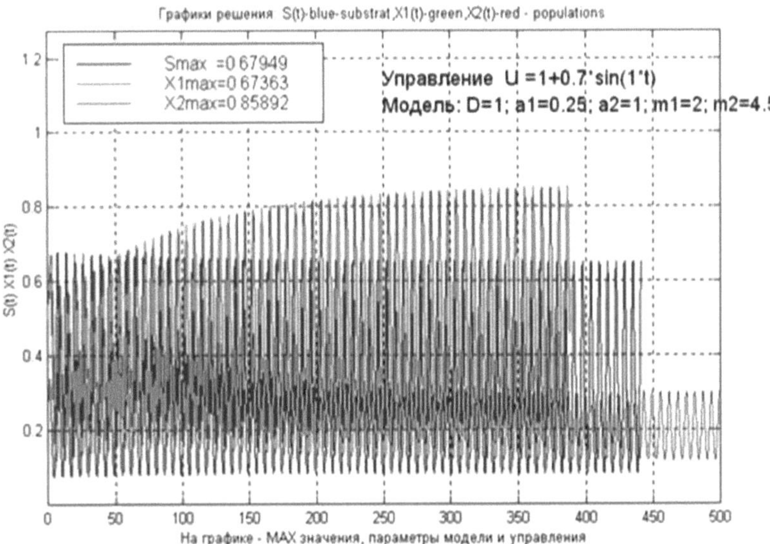

Figure 4.4 - Culture growth character at pulse substrate supply. XI - Lactococcus lactis subsp. lactis X2 - Lactococcus lactis subsp. cremoris. S - substrate. Feeding frequency! substrate 0.075 l/xv. Peryud - 6.25 hv.

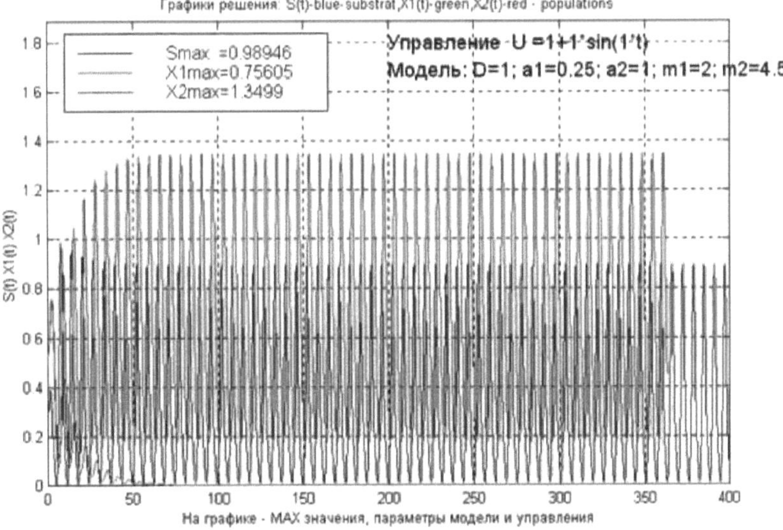

Figure 4.5 - Culture growth character at pulse substrate supply. XI - Lactococcus lactis subsp. lactis X2 - Lactococcus lactis subsp. cremoris. S - substrate. Frequency of substrate feeding 0.28 l/xv. Peryud - 6 xv

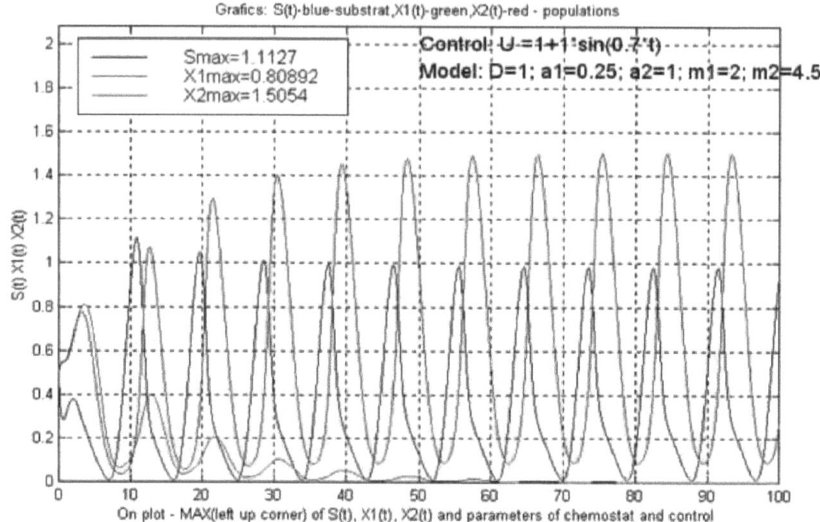

Figure 4.6 - Culture growth character at pulse substrate supply. XI - Lactococcus lactis subsp. lactis X2 - Lactococcus lactis subsp. cremoris. S - substrate. Frequency of substrate feeding 0.2 litres/xv. Peryud - 5 xv

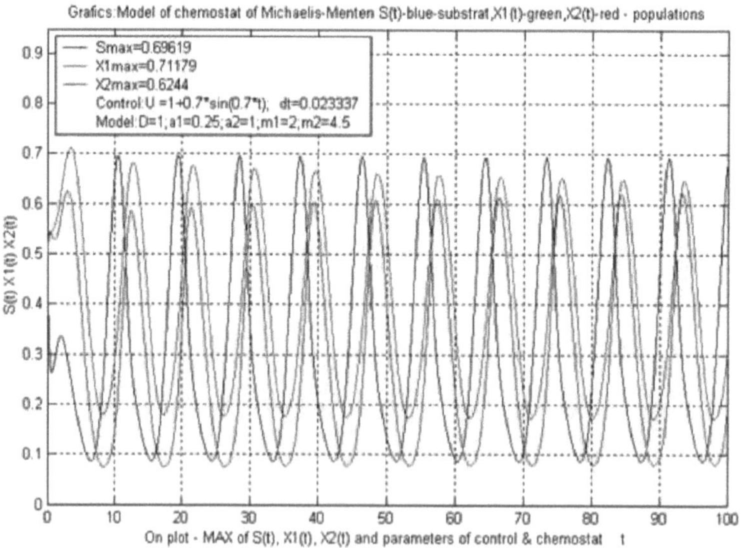

Figure 4.7 - Culture growth character at pulse substrate supply. XI - Lactococcus lactis subsp. lactis X2 - Lactococcus lactis subsp. cremoris. S - substrate. Frequency of substrate feeding 0.06 l/xv. Peryud - 10 xv

CHAPTER 7

RESULTS OF NUMERICAL MODELLING.

The results of numerical modelling are arrays of numbers on which the solution graphs are plotted. The text of the M-file is presented in Appendix 1.

The implementation of numerical methods for solving the Chemostat equations in MATLAB 2011, proposed in this paper, allows us to construct solutions with accuracy sufficient for the needs of practice. Confirmation of the truth of the solutions is the coincidence of the presented curves with the results of analytical calculations of solutions for some special types of input influences. So, for example, transients in the chemostat corresponding to constant input concentrations $So = 0.25$; $X01 = 0.75$; $X02 = 0$, coincide with the analytical results. Similarly, the parameters of the steady-state oscillations for not very large amplitude fluctuations of the input substrate concentration, presented in the graphs in Appendices 2-3, coincide with the parameters of the oscillatory modes, which are calculated by the harmonic linearisation method;

2. The proposed procedure of numerical solution allowed to carry out studies of transients and constant oscillatory modes for ultra-low frequencies and any amplitudes of input influence, when no analytical methods of calculation (including the method of harmonic linearisation) are unsuitable;

3. The results of numerical modelling of processes in the chemostat have shown that by changing the parameters of fluctuations of the input substrate flow it is possible to control the processes of survival and extinction of different populations inside the chemostat; in particular, for the case of two populations - the population dies out at a constant concentration of the input substrate flow, at periodic input influence can have a density greater than the stationary population density, survives at a constant flow this fact can be of great practical importance.

list of sources

1. A. D. Bazykin. Mathematical biophysics of interacting populations. M., Nauka, 1985;

2. P. A. Poluektov et al. Dynamic models of ecological systems. L.: Gidrometeoizdat, 1980;

3. B.G. Zaslavsky, R.A. Poluektov. Management of ecological systems. Moscow: Nauka, 1988;

4. Hsu S. B., Hubbell S., Waltman P. A mathematical theory for singlenutrient competition in continuous cultures of microorganisms, SAM S. S. Appl. Math. 32, 1977;

5. Bailey J. E., Ollis D. F. Biochemical engineering fundamentals. N. Y.: McGraw-Hill, 1977;

6. Butler G. J., Hsu S. B., Waltman P. A mathematical model of the chemostat with periodic washout rate // SIAM J. Appl. Appl. Math. 1985. V. 45. P. 435449;

7. Butler G. J., Wolkowitcz G. S. K. A mathematical model of the chemostat with a general class of functions describing nutrient uptake // SIAM J. Appl. Appl. Math. 1985. V. 45. P. 138- 150;

8. Smith Hal L. Competitive co-existence in an oscillating chemostat // SIAM J. Appl. Appl. Math. 1981. V. 40. No. 1. P. 498-522;

9. Afanas'ev V. N., Kolmanovskii V. B., Nosov V. R. Mathematical theory of conrol systems design. Dordrecht: Kluwer Acad. Publ., 1996;

10. Smith H.L., Waltman P. The theory of chemostat: dynamics of microbial competition. Cambridge University Press, 1995;

11. G.E. Kolosov, D.V. Nezhemetdinova. Journal of Automation and Telemechanics No.1, 2000 "Investigation of steady-state oscillatory processes in chemostat", 2000;

12. Mixed-Culture Fermentations Clifford W. Hesseltine

13. Methodical guidelines for economic substantiation of scientific and research work in the diploma project (for x!м1чnyh specialities) / Compiled by. Karetshkova V.S., Bisher D.N. - X.: NTU XSh. - 2002 - 64 p. The Law of Ukraine "About the protection of forests", leaffall 2002 p.

14. M.P. Benko, I.A. Belikh. Methodical instructions to the performance of the test work at the bachelor's degree, cneuianicTa (change and design of the explanatory note) for students of speciality 6.0929, 7.0929 "Industrial engineering". - NTU "HPI", 2008.

CHAPTER 9

APPENDIX 1

Programme code written in system! MATLAB.

```matlab
function [y]=Chemos(U0, Um, w, So, X1o, X2o, D, a1, a2, m1, m2, T0, Tm);
% function Chemostat_model(U0, Um, w, So, X1o, X2o, D, a1, a2, m1, m2, T0, Tm)
%   modeling of controled chemostat described of system ODE by Michaelis-Menten
% function has 13 positive parameters:
%  U0, Um, w - components of control function U = U0 + Um*sin(w*t)
%  So > 0, X1o > 0, X2o > 0 - inital conditions of system ODE
%  D, a1, a2, m1, m2 - parameters of chemostat
%  T0,Tm - define time span [T0 Tm] of modeling
%if nargin < 1
% U0=1; Um=0.5; w=1; D=1; a1=0.25; a2=1; m1=2; m2=4.5;T0=0; Tm=10;
%end
tspan = [T0; Tm]; y0 = [So X1o X2o];
%check(U0,Um,w,D,a1,a2,m1,m2);
%options     =     odeset('RelTol',1e-8,'AbsTol',[1e-8     1e-8     1e-8],'OutputFcn',@odephas3); фазовый
%портрет
options = odeset('RelTol',1e-8,'AbsTol',[1e-8 1e-8 1e-8]);
[t y] = ode45(@cheMehMenten,tspan,y0,options,U0,Um,w,D,a1, a2, m1, m2);
%-------------------------------------------------------------------------
figure;
%plot(t,y(:,1),'-',t,y(:,2),'-',t,y(:,3),'-',t,control_U,'--');
plot(t,y(:,1),'-',t,y(:,2),'-',t,y(:,3),'-');
zoom on; plotedit on;
solution = [ t , y ]; size_of_solution = size(solution);
dt = (Tm - T0)/size_of_solution(1); delta_t = num2str(dt);
d_t = '; dt=';
```

```matlab
M=max(y);
 d = 'Model:D='; D_chem = num2str(D); apar1=';a1=';
a_1 = num2str(a1); apar2 = ';a2= ';
a_2 = num2str(a2); mpar1 = ';m1= '; m_1 = num2str(m1); mpar2 = ';m2= ';
m_2 = num2str(m2);
parchemostat = strcat(d,D_chem,apar1,a_1,apar2,a_2,mpar1,m_1,mpar2,m_2);
u = 'Control:U =';
Uo = num2str(U0); plus = '+';
um = num2str(Um);
sinus = '*sin('; w_=num2str(w);closeU = '*t)';
control           =           strcat(u,Uo,plus,um,sinus,w_,closeU,d_t,delta_t);
Smax=(max(y(:,1)));
X1max=max(y(:,2));
X2max=max(y(:,3)); Ycoord=[Smax X1max X2max ];
title('Grafics:Model   of   chemostat   of   Michaelis-Menten   S(t)-blue-
substrat,X1(t)-green,X2(t)-red - populations');
xlabel( 'On plot - MAX of S(t), X1(t), X2(t) and parameters of control &
chemostat   t' ); ylabel('S(t) X1(t) X2(t)');
grid;
S    =    'Smax=';    Smax=num2str(Smax);    S_leg    =    strcat(S,Smax);
X1max=num2str(X1max); X1 ='X1max=';
X1_leg = strcat(X1,X1max); X2max=num2str(X2max);
 X2 ='X2max=';
X2_leg = strcat(X2,X2max);
legend(S_leg,X1_leg,X2_leg,control,parchemostat,2);   axis( [ 0,  Tm,  0,
max(Ycoord)+ max(Ycoord)/3] );
% -------------------------------------------------------------------------
function dydt = cheMehMenten(t,y,U0, Um, w, D, a1, a2, m1, m2)
U=U0+Um*sin(w*t);
dydt = [
```

```
(U-y(1))*D-(m1*y(1)*y(2))/(a1+y(1))-(m2*y(3)*y(1))/(a2+y(1))
((m1*y(1))/(a1+y(1))-D)*y(2)
((m2* y(1))/(a2+y(1))-D)* y(3)
];
```

I want morebooks!

Buy your books fast and straightforward online - at one of world's fastest growing online book stores! Environmentally sound due to Print-on-Demand technologies.

Buy your books online at
www.morebooks.shop

Kaufen Sie Ihre Bücher schnell und unkompliziert online – auf einer der am schnellsten wachsenden Buchhandelsplattformen weltweit! Dank Print-On-Demand umwelt- und ressourcenschonend produziert.

Bücher schneller online kaufen
www.morebooks.shop

info@omniscriptum.com
www.omniscriptum.com

Printed by Books on Demand GmbH, Norderstedt / Germany